What Can Math Do For You?

129 Quick Answers

Logan Schoolcraft

DEDICATION

To Math

Hi. My name is Logan and I owe a lot to math. I've claimed it's helped me get my foot in the door for jobs, new experiences, and more things than I can imagine. I'm not the best math person in the world. Far from it. I don't even have an advanced degree. I sometimes struggle with basic math like everyone else; I'm human. But I do know that I would not be where I am today if I did not have the math skills, experience, practice, and understanding that I do today.

So that's where this little book (if you can even call it a book) comes in. I simply had the question, "what can math do for you?" I was posing it to myself mostly and how I would answer it to others. Just what can I do, and what have I done, with math. I was curious of how much math actually touches my life.

I know that math was involved in a lot of stuff I do, but just about everything I could think of was related to math. Or had a math component. Or made me remember a math lesson from years ago. I didn't know if I could separate math out of the things I do.

So I decided to make a list. Most of this list comes from my experiences. But there are some that other people have offered. I wanted to make a big list, but also a readable list. And one that was motivating and encouraging to young students as well as older folks who felt they have struggled with math. A list that could help people see the opportunity and power math holds for them was another criteria.

The result? A list of 129 answers that made me go,

"wow, that's a lot math can do." There's no real sequence to the list either…it's just kind of how things came to me.

No, I don't know all there is to math, but I do know that if I can help one person get a better understanding, then I've succeeded with this list.

So, what can math do for you? …

1

What can math do for you?

It can help you add up your groceries. Ever been on vacation, go into the store only carrying cash? Well, what do you do if you can't spend more than $20? Get to adding. Simple. Easy. Basic. You don't even have to get exact. Try rounding everything up to the next dollar and keep a running total in your head. That way, you should have plenty to cover taxes, or any minor error you may have had.

2
What can math do for you?

Help you figure out how many pieces of gum are in a carton, based on how many are in one packet. It's basically a multiplication problem (five pieces per packet times 10 packets per carton).

3

What can math do for you?

Add up your golf score…unless you want claim you turned professional today.

4

What can math do for you?

Count stuff—like pillows on your bed, friends at a
wedding, cars on the freeway, or stuff you can afford at
the store. Ahh counting. Such a hard thing to do. Be it
counting the broken cookies in your bag, or something
harder—like moving people in the mall. Somehow this
basic task can be really hard, like counting the scratches
on an old windshield. Wow! Never fear though, once
mastered, the skill is one you'll use almost everywhere
you go.

5

What can math do for you?

Math can help you solve how much buttermilk goes in your biscuits when you triple the recipe. 3X= what? No, don't worry, I'm not asking. There are a lot of ways to figure out the amount of this creamy flavor enhancer to put into a larger recipe. Yes, you could algebraically figure it out. You could add the amount three times, or you could just do the recipe three times in a row (in one big bowl). No matter how you do it, math still gets you there.

6

What can math do for you?

Tell you how much your cinnamon rolls cost to make (in
material and time). Yes, at one time, I was considering a
food truck that only made cinnamon rolls, so figuring
out how much each roll would cost me was pretty
important. (That's some tasty math!)

7

What can math do for you?

Let you figure the square footage of your house and its value. Looking to sell, rent, or re-finance, it still works the same.

8

What can math do for you?

Tell you how many pints of beer your British friend drank out of your keg in terms of ounces. (Friend?...)

9

What can math do for you?

Let you understand the wear pattern on your tires. I drive an old truck, and knowing which way the wear pattern is going lets me know the correction to apply; and save money on tires too!

10

What can math do for you?

Create a life-size human maze. Ooooo. I really like this one. Ever built a life-sized maze before? Me neither. Until my mom got the wild idea to make one. (Maybe I was part of the reason she did it....) This one is a lot of fun to me. You have to use several math skills to make things come together; such as geometry, algebra, and addition. When you're done though, the reward is so much fun! (Even for a really big kid like me!)

11

What can math do for you?

Show you how to figure the length to cut a board when making a bird house roof.

12
What can math do for you?

Understand how to apply similarity, and not just to triangles. Side-Angle-Side, Angle-Angle-Angle, what?! Don't worry, this is just similarity between triangles, but math with let you learn the premise here, and be able to take it further.

13

What can math do for you?

Understand what periodic means. No, not the dot at the end of this sentence. I'm talking about repeating patterns and functions. There are a lot of periodic themes and things in our life (just look at the day—the sun comes up and goes down regularly doesn't it?). When you're able to see something occurring periodically, you can make it work for you. Say you see rush hour traffic happens every day at 5pm. You could just shift your work/school/routine to avoid this (if it gives you a headache).

14
What can math do for you?

Show you how circles relate to triangles and area. Yeah, they do!

15

What can math do for you?

Figure a rate for mowing the lawn and how long it would take to do it again. Can you mow a lawn in 2 hours? What about a back yard? Or a pasture? If you know the size of each, and how long it takes you do one, then you can easily figure the answer out. And if you feel enterprising, you can charge based on the rate of your job(s).

16

What can math do for you?

Understand the dry time of paint and know when to apply the second (or when you'll be done). Add to that the rate of flash time, catalyst delay, and more, and you'll be a professional painter in no time flat.

17
What can math do for you?

Make you look like a real nerd when you know how to total up a growth of bacteria. Yeah, it's using a bit of calculus, and no, I haven't done it.

18

What can math do for you?

Tell you what a probability of 1 means and how much you should bet on it.

19
What can math do for you?

It will let you know what batting .333 really means and what's likely when your team is at bat. I loved baseball growing up and at one point had most of the rule book memorized (yeah, maybe I am a little nuts). But I also knew what my stats meant, and my favorite player's. Knowing that a .333 batting average means I hit the ball every third time at bat is useful (even as sad as it sounds). When the World Series in on the line, then you want to know what every stat, in every situation, means.

20

What can math do for you?

Impress your friends by figuring out the day of the week they were born. I had this trick down once, but it's hard to keep unless you use it regularly.

21
What can math do for you?

Guestimate how much water and food you'll need on a
hike, given the season. I ran out of water once on a
hike, and will always add a contingency in now.

22
What can math do for you?

Remember when, what channel number, and how long your favorite tv or radio show is. My brother is a tv number and time nut. I don't know how he knows all those numbers going off in his head. Well, I always thought he was better at math than I was, so maybe that's the skill I missed (but I don't mind).

23

What can math do for you?

It can get you a job. And then help you count your money.

24
What can math do for you?

It can assist you in starting your own business. And then
deciding if it's profitable, sellable, or needs work.

25
What can math do for you?

Help you quit your job. Yep, I did this. I knew I needed a certain amount of money to do it first, and math let me. (And yes, I did feel better after leaving.)

26

What can math do for you?

Cook thanksgiving dinner. Not literally, but almost. Think of all the planning that goes on in that feast—for me, that's the biggest part math helps out with—planning the preparation. Which means what goes in the oven when; for how long; with what else; and when do I change temperatures and foods.

27
What can math do for you?

It can help you cook pretty much anything: veggies, meat, bread (rolls, loaf, cinnamon rolls), snacks, and even drinks and smoothies.

28
What can math do for you?

Make your own beer or wine. This takes some chemistry too, so you'll be doing some cross-disciplinary work. Make the perfect cheese, sourdough bread, mead, wine, beer. This was mentioned earlier, but all these things ferment, so it's a bit of a different process, so you'll have to alter your calculations just a bit.

29
What can math do for you?

It won't let you over pay, at restaurants, clubs, or any other place taking money.

30
What can math do for you?

Gauge your health and fitness. With math, you can make your own spreadsheet of foods and nutrients, learn and understand your body's stats (blood pressure, weight, height, BMI, etc.), and know what to do if any one of them changes.

31
What can math do for you?

Figure a tip. Even make one like 15% seem easy in your head; no joke!

32
What can math do for you?

Plan and plant a garden.

33
What can math do for you?

Help you work on aquaponics (or dream about it like me). FYI, aquaponics is the raising of fish and vegetables in a single closed-loop system, so you need to know a bit about volume, flow, time, and how it all relates together (i.e. functions).

34
What can math do for you?

It can assist you raising chickens (layers, broilers, etc.).
Need to know how much feed to buy? What about a
water consumption rate? No problem with math.

35
What can math do for you?

Feed and exercise lambs for show. An old 4-H project of mine I did with my brother. His math was exercising them, and I did the feed and supplements (just think of it all like bodybuilding).

36
What can math do for you?

Feed a dog. Yep, you better know when Spot has to eat, how much, and if you can afford it, cause you gotta look him in eye…

37
What can math do for you?

Help you figure out how to cook at elevation. Let me just say that cooking beans at elevation is a royal pain unless you know a thing or two (or cheat and get a pressure cooker like I did).

38
What can math do for you?

Tell you how much soda you do or don't drink in a
week. Could be as basic as counting, or as detailed as
the number of sugar grams consumed. Either way,
maybe it will open your eyes.

39
What can math do for you?

Bake a cinnamon roll to perfection. Even though I'm a "sloppy" baker, math still lets me know when these golden beauties are perfect…and it's about the same time my nose does too!

40
What can math do for you?

Show you how to cut the pie so everybody gets equal parts. No whining, either!

41
What can math do for you?

Count calories.

42
What can math do for you?

Aid in scheduling your day so you know the number of things you can realistically get done. I sometimes get the feeling people don't really know how much they can get done, so tabbing it up and actually knowing is a great help. It did wonders for me.

43
What can math do for you?

Track your weight loss.

44
What can math do for you?

Figure your BMI. Pure formula, but I would only use
after you tracked your weight loss.

45
What can math do for you?

Know your lap time when running a marathon, and figure average time per mile. I know a lady who's qualified for the Boston Marathon next year, and she really tracked her times in training leading up to her qualification run—I got the feeling every second counts.

46

What can math do for you?

Help you weld. Believe it or not, it's the prep work (as in getting things cut to fit) that makes a great weld come together. And that's a math skill.

47
What can math do for you?

Solder. Timing, cleanliness, and perfect sizing make
solder work a breeze.

48
What can math do for you?

Understand carpentry work. Here too, making things fit is key—and that's nothing more than geometry.

49
What can math do for you?

It can help you understand screw, nut, bolt, and washer sizes.

50
What can math do for you?

Fix a golf cart. Will the lift-kit your boss purchased
actually clear the fenders? (Where's the tape
measure?...)

51
What can math do for you?

Paint a truck. Not just paint, but plan it, the help, and the entire interaction that goes on.

52
What can math do for you?

Run a sand blaster. Crazy loud, but really effective. But do you know your flow rate? Got a handle on how much air you are using? All this cool process revolves around math.

53
What can math do for you?

Help you assemble a metal building. Got your footing right? Are the bolt holes spaced correctly? Where did the drill go?... All aspects of construction bring out your math skills.

54
What can math do for you?

Pour a concreate foundation for your metal building.
You'll feel like there wasn't any math going on when
you're done, but your wallet might be the first place that
does.

55
What can math do for you?

Build a latter/step stool. Angles, lengths, screws and some cuts. Simple, neat, and fun.

56
What can math do for you?

Help you build a tiny house. Or at least sketch out how
much material you'll need to frame it up, and then
realize you now have one BIG project on your hands.
(Really fun though.)

57
What can math do for you?

Fix your truck's brakes. Or timing. Or air/fuel ratio, etc., etc. My old truck is nothing but a math lesson for me.

58
What can math do for you?

Repair a water hose.

59
What can math do for you?

Build a water system. You might not stop to think how many parts go into a water system. (For the high purity one I built, it was a softener, backwash filter, 2 pre-filters, RO system (with its own filters), circulation tank, deionization pods, pump, UV lamp, and water heater. Those were the big components anyway. Getting everything together was the hard part…and not wasting time or money.)

60
What can math do for you?

Powder coat (almost anything). My brother used some cool geometry to build his oven for this—didn't waste a single piece of material. And he kind of does the same for his powder coating too. (Could he really be better than I am at a lot of stuff?...)

61
What can math do for you?

Help you run a lathe. Need to take .25" off that shaft?
Don't forget you only need to move your cutter in half
that amount (diameter vs. radius).

62

What can math do for you?

Run a mill. Where is center on this thing? (A question I seemed to ask a lot). Figuring out how to use your cutting bit to center is just one way of algebraically figuring your way to an answer.

63
What can math do for you?

Help with building, or converting, a vehicle to electric. This is still just a dream of mine, but nevertheless, I still dream in conversion functions (W=VxA).

64
What can math do for you?

Math can help you go solar. My RV only has a 30amp service shore line, so if wanted to go solar, all I have to do is match that. (But when you pay ~$100/watt, I get a little tight with my wallet when the total is in the thousands!)

65
What can math do for you?

Make a shed for the patio during summer. Do you know your latitude? Then you know what the angle of the sun is during the summer. And you would then be able to figure out how much of an overhang you'd need to keep your patio cool and in the shade. I helped do this for a bee-port (a cover for bee hives).

66
What can math do for you?

Math can help you pass your class(es); especially math. But aside from math class, you'll be able to compute your grade at any point in the term, and know if you're going to pass.

67
What can math do for you?

Get you graduated; from high school, college, or any other educational institution.

68
What can math do for you?

Calculate the health of your vehicle. Just think MPG.
This is a simple calculation that, if done routinely, will
tell you a lot about the condition of your car.

69
What can math do for you?

Help you do repairs and maintenance, on your car/house/RV/motorcycle/toys. Heck, an oil change is nothing more than a subtraction and addition problem.

70
What can math do for you?

Find the best price on goods and services. Without your mathematical skills and understanding, what would comparison shopping be? Example: buying a car. Need I say more?

71
What can math do for you?

Help you balance a checkbook; namely your own. Yes, you can have some fancy percentages going on, but for the most part, it's just addition and subtraction…with your money (yeah, the subtraction part hurts).

72
What can math do for you?

Estimate time. Be it for a hike, trip/travel (driving), or time off work.

73
What can math do for you?

It lets you use and understand a calculator. If you have no math in your memory, what good is a calculator? You still have to know what the buttons do, and what to enter, and in what order.

74
What can math do for you?

Create and understand spreadsheets. Accountants love them, a lot of other people don't. But you can gain a lot of information from them, in a little time. Love sports and want to compare all your favorite players, side by side? A spreadsheet can do it.

75

What can math do for you?

Help you write a book. Kind of like this one. I have to know how many pages, at a given size, I need for a total number of words. Then I have to be able to organize things in a logical, or understandable manner. All math concepts.

76
What can math do for you?

Read a book. I'm reading a thick one right now, and I'm constantly figuring out my rate, and how much longer I have to read it. (Don't get me wrong, I love reading, but sometimes long books should be made into smaller ones.)

77
What can math do for you?

Figure your water/electricity/gas bill.

78
What can math do for you?

Plan for the weather. If it's going to be 10 degrees outside, do you take your jacket, or stay home? Depends. Are we talking Fahrenheit or Celsius?

79
What can math do for you?

Impress your friends. Mental math is only one way…

80

What can math do for you?

Make sure your inventory of library and book fair books is correct.

81
What can math do for you?

Create, edit, produce, and sell a video on one of your
passions. There are so many math concepts going into
this one that I'd need a bit more space to flesh it out.

82
What can math do for you?

Dispense medication (for yourself and those you love….even doggie).

83
What can math do for you?

Budget your monthly expenses.

84
What can math do for you?

Play the cello (or any other instrument). What is an
octave? What resonance makes a note out of tune?
Understanding division (2, 4, 8, etc.) will help you get
the gist of the rhythms as well. Play on!

85
What can math do for you?

Tell time.

86
What can math do for you?

Measure time.

87
What can math do for you?

Conduct an experiment. But please be nice to your little
brother…

88
What can math do for you?

Make sure you understand order of magnitude. Did the cost come to $100 or $1,000? Here, the decimal point really matters; especially when it's my money.

89
What can math do for you?

Gamble (lottery, dice, cards, horses, etc.) It's all probability.

90
What can math do for you?

Figure the water bill for the swimming pool you've always wanted. If you're in the desert, please don't get one though; there are better uses for that much water. Just compute how many people you can provide water for before doing it—it will astound you!

91
What can math do for you?

Estimate how much honey your bees will produce. A rather sweet problem!

92
What can math do for you?

Find the optimal production rate to optimize profit in your small business.

93
What can math do for you?

Sail a boat.

94
What can math do for you?

SCUBA dive. Another multidisciplinary study, but so much fun. Who knew handling the pressure would be that much fun?

95
What can math do for you?

Sky dive. Calculus in full speed.

96
What can math do for you?

Mail a Christmas gift, so nobody gets left out and you don't go broke.

97
What can math do for you?

Handle a credit card.

98
What can math do for you?

Count and understand money—the bills and coins.

99
What can math do for you?

Sharpen a knife. What is the difference between a five degree angle and a fifty degree angle, other than an order of magnitude? With a knife, it's the difference in sharp and dull!

100
What can math do for you?

Gauge your improvement (personally, professionally, academically, or any other discrete way).

101
What can math do for you?

Clean your house.

102
What can math do for you?

Clean a large optic. Guilty! I know how much it costs
to the cent in labor and parts, as well as how much
material and acid is needed (and for how long) to clean a
segment of the Hobby Eberly Telescope. I did it for far
too long…

103
What can math do for you?

Know how to buy a computer. Or tablet. Or video camera. Or any electronic device that deals with MB, GB, pixels, MHz, GHz, and the like.

104
What can math do for you?

Understand your phone and its bill. 20 years ago this would not apply, but now, it's almost a must; as my mother knows.

105
What can math do for you?

Program a PLC or computer.

106
What can math do for you?

Help you learn and apply logic. In your daily life, and even more structured, as in your writing, or argumentation skills.

107
What can math do for you?

Improve your problem solving skills.

108
What can math do for you?

Help you change a water fitting on your RV. You better know the size of the inlet and the outlet. And how long it will last. (So how long has the current one been in there?...)

109
What can math do for you?

Figure the slope of your hike up Guadalupe Peak, in
Guadalupe Mountains National Park.

110
What can math do for you?

Teach you about negatives—like losing money, going in reverse, or jumping off a diving board (more gravity).

111
What can math do for you?

Understand wind chill, as bad as it sounds.

112
What can math do for you?

Help you teach.

113
What can math do for you?

Help you tutor. Especially math!

114
What can math do for you?

Help you save money…or spend it!

115
What can math do for you?

Know how much you'll have at retirement.

116

What can math do for you?

Let you know how long it will take mold to cover your basement, if you don't fix that leaky water valve.

117
What can math do for you?

Tell you how many ways you can (go to work, wear your clothes, make a sandwich, etc.) It's just a matter of combinations and permutations when getting dressed, really.

118
What can math do for you?

Make a list of things you can do with math—like this book.

119

What can math do for you?

Find angles to build something cool. Like a frame for
your certificate from the state!

120
What can math do for you?

Figure measurements to decorate your house. Not for me, honestly!

121
What can math do for you?

Calculate your latitude and longitude to know where you are. Really tough though, if you were just kidnapped and thrown in the bottom of a ship for 15 hours…

122
What can math do for you?

Figure out how much longer you have on your phone battery (and/or plan minutes). I just hate it when I have that figured out, and I hit a dead spot, and my phone just loses battery searching for service.

123

What can math do for you?

Figure out how much weight and length of rope is needed when doing a rescue, or just rock climbing. Yes, I have been on a rope rescue, and no, you can never have enough rope or gear. That's the reason I ended up on the rescue; the cliff was 100 feet down, but the initial crew only had two 150 foot ropes.

124
What can math do for you?

Make a mask for the Hobby Eberly Telescope coating chamber. Just trust me, this does exist, and I had to dig pretty hard in my math memory to work on it.

125
What can math do for you?

Tell you a stylus profiler's standard deviations, standard error, and more, so you know the reliability of your data and equipment. (More coating data…)

126
What can math do for you?

Help you see trends or random chances in stuff—
statistics (like grackle behavior). Just watch the birds
some time and see if you see things coming together. Is
it real, or imagined? Could you graph it? Or predict the
next move?

127
What can math do for you?

Size a truck to haul your RV.

128
What can math do for you?

Math can help you sew (your clothes, a quilt, hat, etc.).
This is cool—geometry, algebra, basic math, all rolled
into one. And you get some really cool duds to wear in
the end too! A complete win-win!

129
What can math do for you?

Tell you how long your bottle of soap will last. Unless you have visitors, of course.

ABOUT THE AUTHOR

Logan loves reading, cooking, and doing math that helps him figure out his next "project".